Harper Hedgehog is a detective on a top sec[ret]
Mayor of Spring Falls, who is called Mighty M[oose]
trying to find Mighty Moose for a special assignment, but hasn't been
able to yet. Harper has decided to ask the animals of Spring Falls to
learn more about the Mayor and to find out where he is.

Harper needs your help, though! Can you help her find Mighty Moose
and then learn what his special assignment is?

Here are your instructions:

🍄 Read the story.

🍄 Trace the words.

🍄 Learn the nouns.

🍄 Solve the mystery word puzzle at the end of the book to find out Mighty Moose's assignment!

My name is HarperHedgehog and I'm new to the town of SpringFalls.

What's your name? I like your pink ears.

Hi, I'm Frankie Fox. I like your purple spines !

Thanks! I'm here to meet friends in SpringFalls.

That's a good idea. We can be friends!

Great! Do you know other animals in Spring Falls?

I sure do! Have you met Mayor Mighty Moose yet?

Not yet. I am looking for him to help me with a problem. Is he your friend too?

He's the best! When I came to Spring Falls, he helped move my TV into my den.

Wow, he must be strong!

He is so strong! He is the strongest animal in town.

That's amazing! I want to meet Mayor Mighty Moose.

I'm busy now, but go find Rocco Racoon. He will take you to meet the Mayor!

Thanks! I'll go find Rocco now.

Hello, are you Rocco ?

Yes I am! Welcome to the Racoon residence.

BOX TOOLS

16 17

My name is _Harper_. I love your blue _tail_ ! _Frankie Fox_ told me I should find you. She said you can take me to meet the _Mayor_.

So you want to meet Mayor Mighty Moose? That's a good idea. He's a great friend to have in this town.

BOX TOOLS

16 17

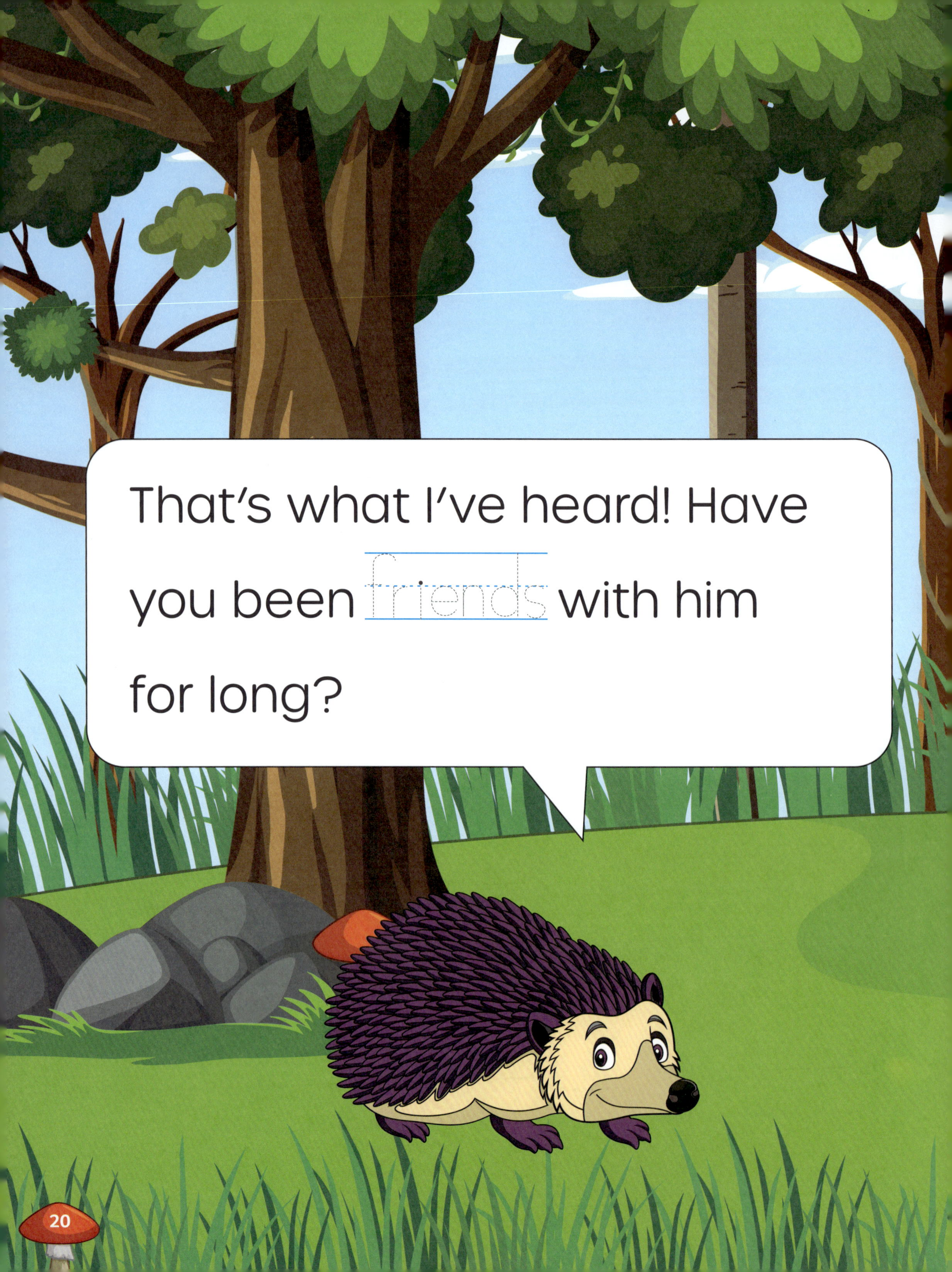
That's what I've heard! Have you been friends with him for long?

Oh yes! When I moved here a long time ago, the Mayor helped me start my car repair shop.

Wow, the more I hear about him, the better he sounds!

You've got that right! Since the Mayor helped me, I'm so busy that I almost never leave the shop.

I can't wait to meet him! Can you take me to him now?

I'm sorry, kid.

I've got to finish the repairs on Sammy Squirrel's car.

It's acorn season, so he really needs his car back.

I guess I can wait until another day .

Hold on! I have an idea. I have a friend who may be able to take you today. His name is Ollie Owl. He's a hoot. You can find him in the tallest tree in town.

That's great! I'll go find Ollie now. It was great to meet you, Rocco !

Great to meet you too, kid .
If you ever need a car repair ,
I'm your guy . I'll give you a
great deal !

29

Hello, my name is Harper.

I like your pink wings!

What's your name?

Hi, I'm Ollie Owl. Good to meet you, dude !

Rocco Racoon told me to find you.

Awesome! Rocco is a cool guy .
He helps me fix my skateboard
sometimes.

33

Do you like to skateboard ?

Of course, _dude_ ! I like to fly, but riding a _skateboard_ is so fun!

Maybe you can show me a _trick_ sometime.

Anytime! What brings you to my tree ?

I keep hearing about Mayor Mighty Moose and I want to meet him. I heard that you can take me to his place .

So you want to meet the Mayor ! That's exciting!

He is the coolest dude . He taught me how to ride a skateboard !

That's amazing! The Mayor sounds like the best friend anyone could ask for.

He really is. Come on, let's go to his house . Let's find Mayor Mighty Moose !

Is there anything else I should know about the Mayor ?

There's lots of stuff! He's a great skydiver. He's never had a cavity. He can play guitar really well. He's a five time Spring Falls Hotdog Eating Contest Champion!

43

That's the most amazing thing

I've heard about Mayor Mighty

Moose all day !

Well get ready to meet the Mayor himself! Here is his house ! I'll knock on the door .

Ollie Owl! I'm so happy to see you!

Mayor Mighty Moose! Good to see you, dude!

Have you learned any new tricks on your skateboard?

You know it, _Mayor_! The reason I'm here is to introduce you to _Harper Hedgehog_.

49

I'm so happy to finally meet you, Mayor !

It's great to meet you, Harper.
Your spines are my favorite
shade of purple !

51

Thank you. I've been trying to find you all day.

Oh really? Can I help you with something?

54

Ok! What do you need? A ride to the airport? Tips on table tennis? A heavy rock moved?

No, Mayor. This is much more serious.

Oh, I see.

Mayor , I have a a top secret special assignment for you.

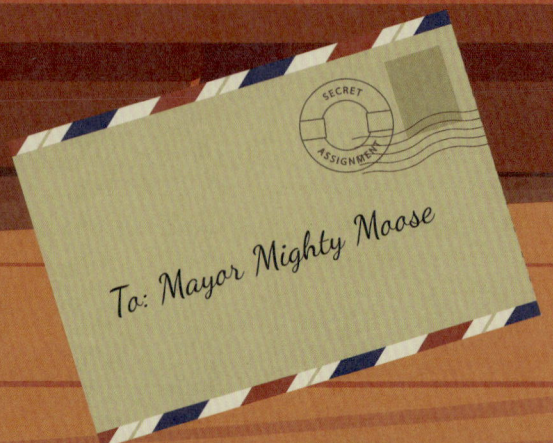

Do you accept?

To: Mayor Mighty Moose